A SIMPLE APPROACH TO SURD

A Step-by-Step Review and practice workbook with exercises and multiple examples on Rational and Irrational Numbers, Operations, Conjugate, Rationalising the denominator, Power, Exponent, Square root, and Equations of Surds.

Adegboye Samuel

TABLE OF CONTENTS

Chapter 1

RATIONAL AND IRRATIONAL NUMBERS

WHAT ARE RATIONAL NUMBERS?

Rational numbers are numbers that can be expressed in a ratio of two integers. For example;

<u>Positive numbers</u>

$3 \rightarrow$ *it can be expressed as a fraction, that is in the ratio of two integers, and this will be equal to* $\dfrac{3 \rightarrow integer}{1 \rightarrow integer}$

$\dfrac{7}{3} \rightarrow$ *it is a fraction, that is the ratio of two integers,* $\dfrac{7 \rightarrow integer}{3 \rightarrow integer}.$

Additionally, the equivalent of $\dfrac{7}{3} = \dfrac{14}{6} = \dfrac{21}{9}$ *are rational numbers, because they are expressed in the ratio of two integers.*

<u>Negative numbers</u>

$-4 \rightarrow$ *it is a rational number because it can be expressed as a ratio of two integers, and this will be equal to* $\dfrac{-4 \rightarrow integer}{1 \rightarrow integer}.$

more examples include $\dfrac{-8}{2}, \dfrac{-12}{4}$ *e.t.c.*

<u>Decimal numbers</u>

$2.25 \rightarrow$ *2.25 is a rational number because it can be expressed as a fraction, that is in the ratio of two integers, and this will be equal to* $\dfrac{225 \rightarrow integer}{100 \rightarrow integer} = \dfrac{9}{4}.$

A fraction with a repeated decimal value

$\frac{2}{3} = 0.6667 \rightarrow$ *this is a rational number because it can be expressed as a fraction, that is in the ratio of two integers, and that is to* $\frac{2 \rightarrow integer}{3 \rightarrow integer}$.

Note: *Non-terminated decimal with different values are not rational numbers. You will get to understand better in my later examples.*

With the above examples, you should be able to identify rational numbers.

WHAT ARE IRRATIONAL NUMBERS?

Irrational numbers are numbers that cannot be expressed as a ratio of two integers. Irrational numbers have their number as a decimal that cannot terminate.

Examples;

- ***Pi*** $\rightarrow \pi = 3.142159265 \dots$ *this numbers are continous*

- ***Euler's Number*** $\rightarrow e = 2.71828 \dots$

- ***Golden ratio*** $\rightarrow \varphi = 1.618033988 \dots$

- $\sqrt{2} = 1.4142356 \dots$

- *Square root of all natural numbers **expect** for **perfect square** are irrational numbers.*

- Surds are also irrational numbers.

Chapter 2

INTRODUCTION TO SURDS

Surds are the roots of rational numbers. It's also called irrational numbers. Examples of rational numbers are 5, $^{7}/_{3}$, –3, 1.25 as explained in the previous chapter. While Numbers that cannot be written as a ratio are called irrational or non-rational numbers. Examples π= 3.141592........., $\sqrt{3}, \sqrt{8}$.

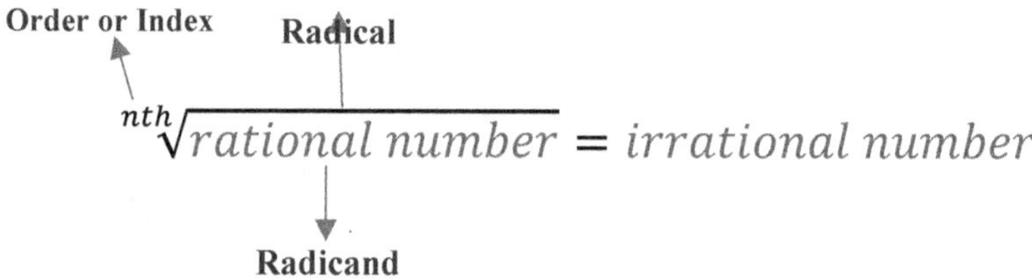

$$\sqrt[nth]{rational\ number} = irrational\ number$$

Examples of surds are $\sqrt{3}$, $\sqrt{14}$, $\sqrt{28}$.

LIKE AND UNLIKE SURD

Like surds are surds that have the same radicands. This means the numbers under the square root sign are the same. Example of like surds include; $\sqrt{3}$, $2\sqrt{3}$, $\frac{5}{2}\sqrt{3}$. While **Unlike surds** are surds that have different radicands, that is, the numbers under the square root sign (radical) are different. Example $\sqrt{3}$, $\sqrt[2]{5}$, $\sqrt[4]{17}$.

ADDITION AND SUBTRACTION OF LIKE AND UNLIKE SURDS.

Like surds can be added or subtracted. examples are shown below.

Examples

1.

$$3\sqrt{2} \quad + \quad 5\sqrt{2} = \quad \mathbf{8\sqrt{2}}$$

$$3+5 \qquad \sqrt{2} \qquad\qquad 8\sqrt{2}$$

2.

$$7\sqrt{3} \quad - \quad 5\sqrt{3} = \quad \mathbf{2\sqrt{3}}$$

$$7-5 \qquad \sqrt{3} \qquad\qquad 2\sqrt{3}$$

3.

$$\frac{1}{2}\sqrt{7} \quad + \quad \frac{1}{2}\sqrt{7} = \quad \mathbf{1\sqrt{7}}$$

$$\frac{1}{2}+\frac{1}{2} \qquad \sqrt{7} \qquad\qquad 1\sqrt{7}$$

4.

$$9\sqrt{5} \quad - \quad 3\sqrt{5} \quad + \quad 2\sqrt{5} \quad = \quad \mathbf{8\sqrt{5}}$$

$$9-3+2 \qquad \sqrt{5} \qquad\qquad 8\sqrt{5}$$

Unlike surds, the addition or subtraction of unlike terms remains the same value.

Multiplication of Like and unlike surds.

Like surds can be added or subtracted. examples are shown below.

Examples

1.

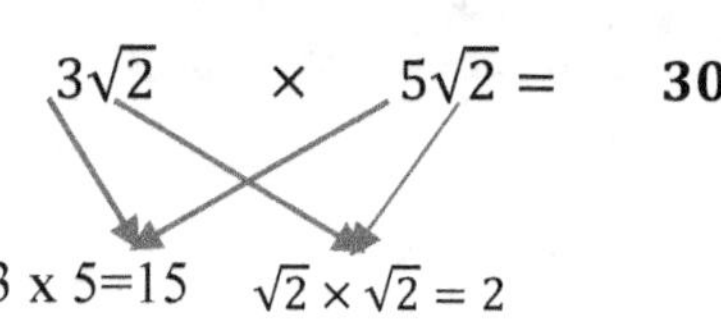

$$3\sqrt{2} \quad \times \quad 5\sqrt{2} = \quad 30$$

$$3 \times 5 = 15 \qquad \sqrt{2} \times \sqrt{2} = 2$$

2.

$$7\sqrt{3} \quad \times \quad 5\sqrt{3} = \quad 35 \times 3 = 105$$

$$7 \times 5 = 35 \qquad \sqrt{3} \times \sqrt{3} = 3$$

For unlike surds;

$$4\sqrt{2} \quad \times \quad 7\sqrt{3} = \quad 28 \times \sqrt{6} = 28\sqrt{6}$$

$$4 \times 7 = 28 \qquad \sqrt{2} \times \sqrt{3} = \sqrt{6}$$

TYPES OF SURDS

SIMPLE SURDS

These are surds that has only one term. For example, $\sqrt{7}, \sqrt{5}, \sqrt{11}$.

PURE SURDS

These are surds that are wholly irrational that is, they have no rational factor. For example, $\sqrt{3}, \sqrt{5},$

SIMILAR SURDS

These are surds that have the same surds factor or they are surds that have their radicands the same. For example, $\sqrt{7}$, $2\sqrt{7}$, $3\sqrt{7}$.

MIXED SURDS

These are surds that have the combination of a whole number and a surd. They are also referred to as surds that have a rational coefficient other than unity. For example, $5\sqrt{2}$, $4\sqrt{3}$, $2\sqrt[3]{5}$.

COMPOUND SURDS

These are expressions that contain the addition or subtraction of two or more surds. For example, $\sqrt{3} + \sqrt{7}$, $3\sqrt{2} - \sqrt{3}$.

BINOMIAL SURDS

These are also called compound surds consisting of two or more surds. They are of the form $a\sqrt{b} + c\sqrt{d}$.

Exercise 2.1

Simplify the following.

1. $8\sqrt{2} + 3\sqrt{2} - 5\sqrt{2}$

2. $3\sqrt{3} + 4\sqrt{3} - 5\sqrt{2} + 2\sqrt{2}$

3. $4\sqrt{5} + 11\sqrt{5}$

4. $7\sqrt{7} + \sqrt{7} + 2\sqrt{7}$

5. $21\sqrt{11} - 4\sqrt{11} - 7\sqrt{11}$

6. $9\sqrt{5} \times 7\sqrt{5} \times 2\sqrt{3}$

7. $\left(2\sqrt{3}\right)^2$

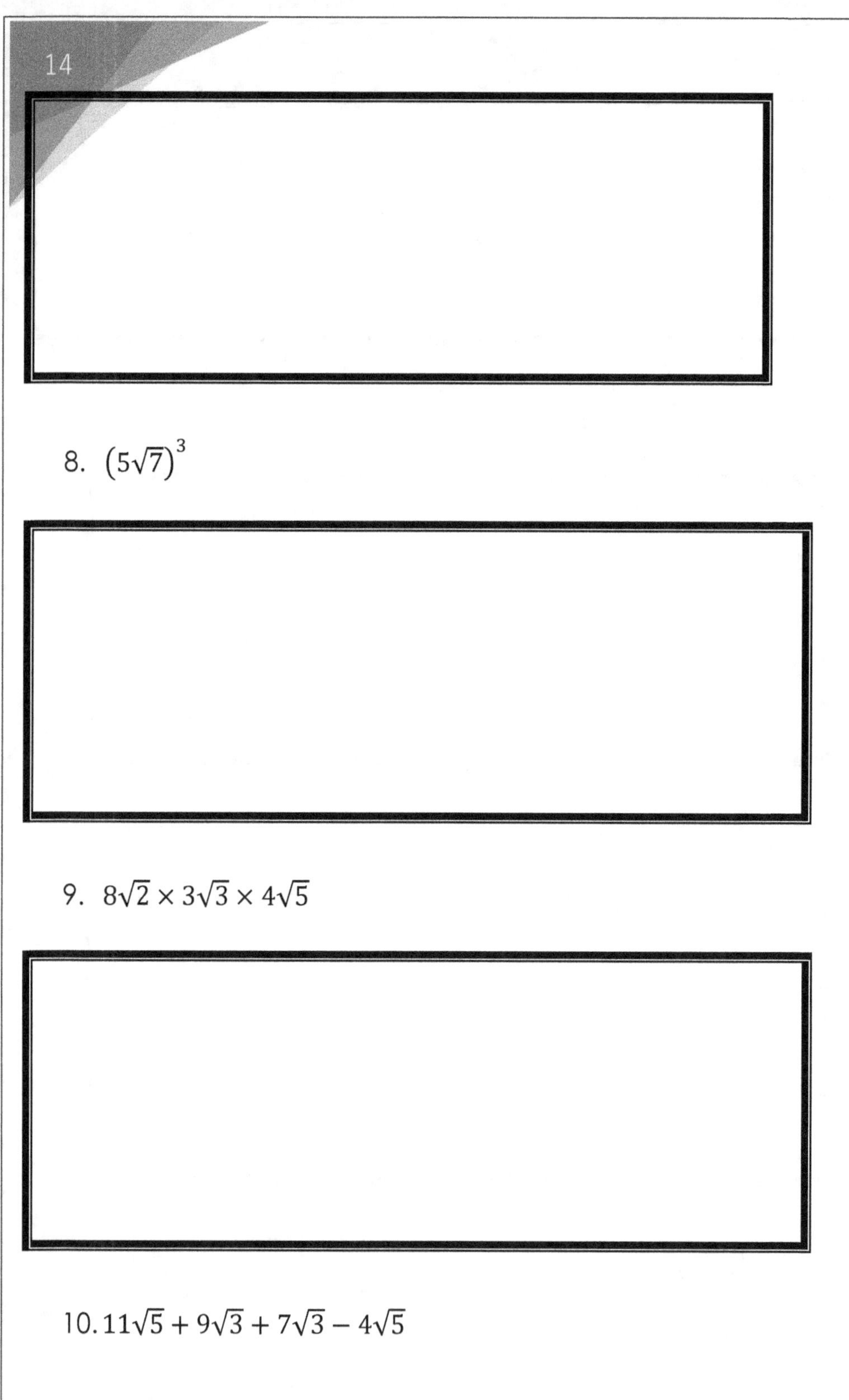

8. $\left(5\sqrt{7}\right)^3$

9. $8\sqrt{2} \times 3\sqrt{3} \times 4\sqrt{5}$

10. $11\sqrt{5} + 9\sqrt{3} + 7\sqrt{3} - 4\sqrt{5}$

Chapter 3

RULES OF SURDS

Some rules govern the evaluation of surds and they are given below;

a. $\sqrt{x \times y} = \sqrt{x} \times \sqrt{y}$

Example

$$\sqrt{16} \times \sqrt{4} = 4 \times 2 = 8$$

Likewise;

$$\sqrt{16} \times \sqrt{4} = \sqrt{16 \times 4} = \sqrt{64} = 8$$

b. $\sqrt{\dfrac{x}{y}} = \dfrac{\sqrt{x}}{\sqrt{y}}$

Example

$$\frac{\sqrt{36}}{\sqrt{9}} = \frac{6}{3} = 2$$

Likewise;

$$\frac{\sqrt{36}}{\sqrt{9}} = \sqrt{\frac{36}{9}} = \sqrt{4} = 2$$

c. $\sqrt{x} + \sqrt{y} = it\ can't be\ simplified$

Example

$$\sqrt{2} + \sqrt{3} = \sqrt{2} + \sqrt{3}$$

In the previous chapter, you were made to understand that, unlike surds cannot be added. It will still give the same value.

This aligns with the third rule that, it cannot be simplified.

d. $\sqrt{x} \times \sqrt{x} = x$

Example

$$\sqrt{2} \times \sqrt{2} = 2$$

e. $\left(x + \sqrt{y}\right)^2 = \left(x + \sqrt{y}\right)\left(x + \sqrt{y}\right) = x^2 + 2x\sqrt{y} + y$

Example

$$\left(2 + \sqrt{3}\right)^2 = \left(2 + \sqrt{3}\right)\left(2 + \sqrt{3}\right)$$

$$2\left(2 + \sqrt{3}\right) + \sqrt{3}\left(2 + \sqrt{3}\right)$$

$$= 4 + 2\sqrt{3} + 2\sqrt{3} + \left(\sqrt{3}\right)^2$$

$$= 4 + 4\sqrt{3} + 3$$

$$= 7 + 4\sqrt{3}$$

f. $\left(x + \sqrt{y}\right)\left(x - \sqrt{y}\right) = x^2 + x\sqrt{y} - x\sqrt{y} - y = x^2 - y$

Example

$$\left(2 + \sqrt{3}\right)\left(2 - \sqrt{3}\right)$$

$$2\left(2 - \sqrt{3}\right) + \sqrt{3}\left(2 - \sqrt{3}\right)$$

$$= 4 - 2\sqrt{3} + 2\sqrt{3} - \left(\sqrt{3}\right)^2$$

$$= 4 - 3 = 1$$

g. $\dfrac{x}{\sqrt{y}} = \dfrac{x}{\sqrt{y}} \times \dfrac{\sqrt{y}}{\sqrt{y}} = \dfrac{x\sqrt{y}}{y}$

Example

$$\frac{2}{\sqrt{3}} = \frac{2}{\sqrt{3}} \times \frac{\sqrt{3}}{\sqrt{3}} = \frac{2\sqrt{3}}{3}$$

The basic rules stated above we help our understanding when it comes to evaluation and simplification of Surds.

Exercise 3.1

Simplify the following.

1. $\sqrt{4 \times 9}$

2. $\sqrt{256 \times 169 \times 144}$

3. $\sqrt{\dfrac{169}{144}}$

4. $\sqrt{\dfrac{36 \times 16}{4 \times 9}}$

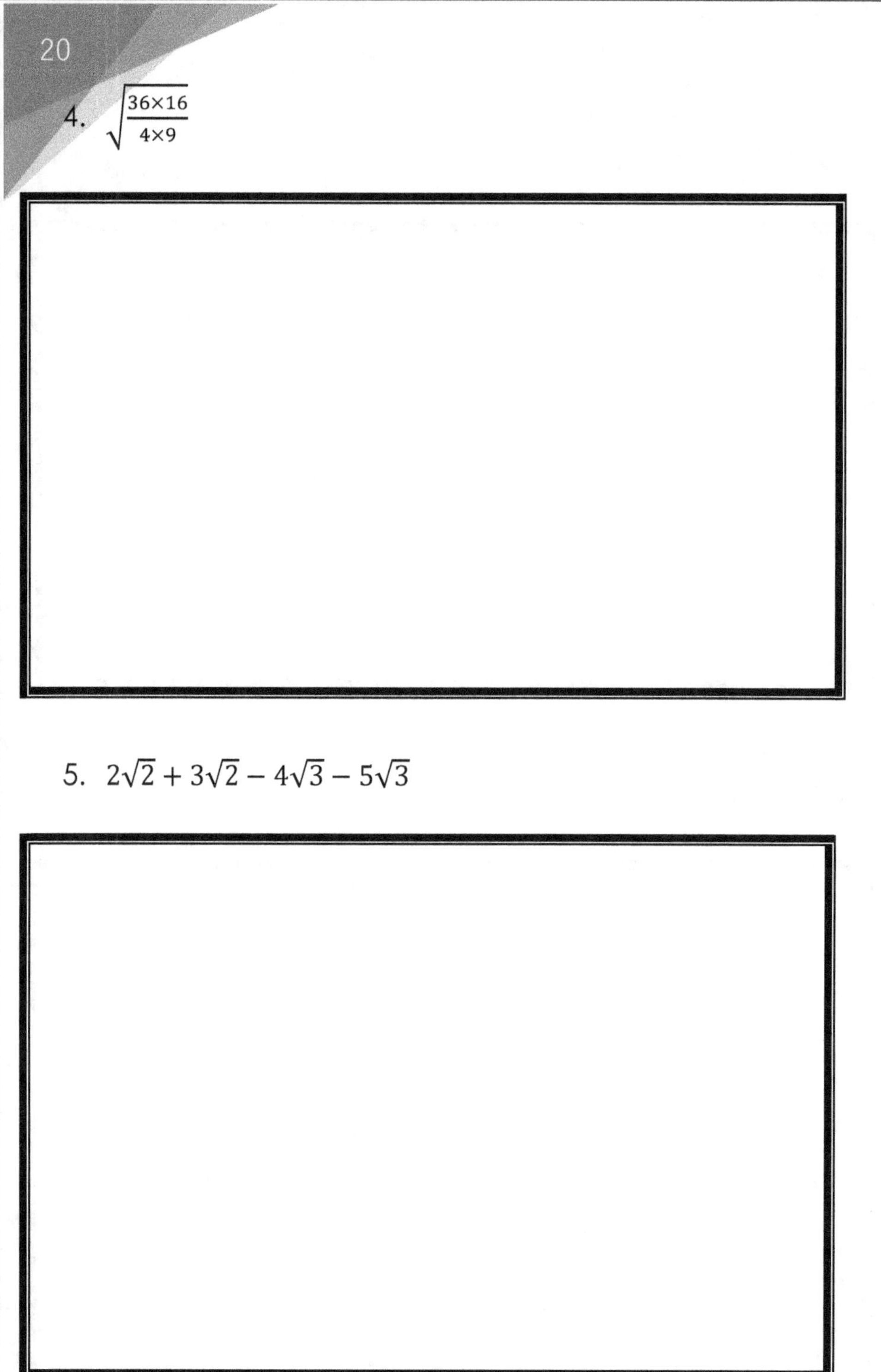

5. $2\sqrt{2} + 3\sqrt{2} - 4\sqrt{3} - 5\sqrt{3}$

6. $\left(2 + \sqrt{5}\right)^2$

7. $\left(\sqrt{3} + \sqrt{7}\right)^2$

8. $(\sqrt{3} - \sqrt{2})(\sqrt{3} + \sqrt{2})$

9. $\sqrt{\dfrac{3}{4}}$

10. $\left(\sqrt{7}\right)^2$

C h a p t e r   4

EVALUATION OF SURDS

SIMPLIFICATION OF A SINGLE SURD TO BASIC FORM

To simplify a single surd

a)   *Express the number under the square-root sign as a product of two factors, of which one is a perfect square.*

b)   *Find the square root of the perfect square*

Question 1: Simplify $\sqrt{45}$

To achieve this, find the factors of 45, such that one is a perfect square.

$$\sqrt{45} = \sqrt{9 \times 5}$$
$$= \sqrt{9} \times \sqrt{5}$$
$$= \sqrt{3 \times 3} \times \sqrt{5}$$
$$= 3 \times \sqrt{5}$$
$$= \mathbf{3\sqrt{5}}$$

Question 2 : Simplify $\sqrt{162}$

To achieve this, find the factors of 162, such that one is a perfect square.

$$\sqrt{162} = \sqrt{81 \times 2}$$
$$= \sqrt{9 \times 9} \times \sqrt{2}$$

$$=\sqrt{9 \times 9} \times \sqrt{2}$$

$$=9 \times \sqrt{2}$$

$$=\mathbf{9\sqrt{2}}$$

Question 3: Simplify $\sqrt{a^2 b}$

To achieve this, find the factors of $a^2 b$, such that one is a perfect square.

$$\sqrt{a^2 b} = \sqrt{a^2 \times b}$$

$$=\sqrt{a^2} \times \sqrt{b}$$

$$=\sqrt{a \times a} \times \sqrt{b}$$

$$=a \times \sqrt{b}$$

$$=\boldsymbol{a\sqrt{b}}$$

Question 4: Simplify $\sqrt{20}$

To achieve this, find the factors of 20, such that one is a perfect square.

$$\sqrt{20} = \sqrt{4 \times 5}$$

$$=\sqrt{4} \times \sqrt{5}$$

$$=\sqrt{2 \times 2} \times \sqrt{5}$$

$$=2 \times \sqrt{5}$$

$$=\mathbf{2\sqrt{5}}$$

Question 5: Simplify $\sqrt{54}$

To achieve this, find the factors of 54, such that one is a perfect square.

$$\sqrt{54} = \sqrt{9 \times 6}$$
$$= \sqrt{9} \times \sqrt{6}$$
$$= \sqrt{3 \times 3} \times \sqrt{6}$$
$$= 3 \times \sqrt{6}$$
$$= 3\sqrt{6}$$

Question 6: Simplify $\sqrt{99}$

To achieve this, find the factors of 99 , such that one is a perfect square.

$$\sqrt{99} = \sqrt{9 \times 11}$$
$$= \sqrt{9} \times \sqrt{11}$$
$$= \sqrt{3 \times 3} \times \sqrt{11}$$
$$= 3 \times \sqrt{11}$$
$$= 3\sqrt{11}$$

Question 7: Simplify $\sqrt{288}$

To achieve this, find the factors of 288 , such that one is a perfect square.

$$\sqrt{288} = \sqrt{144 \times 2}$$
$$= \sqrt{144} \times \sqrt{2}$$
$$= \sqrt{12 \times 12} \times \sqrt{2}$$
$$= 12 \times \sqrt{2}$$

$$=12\sqrt{2}$$

Question 8: Simplify $\sqrt{147}$

To achieve this, find the factors of 288 , such that one is a perfect square.

$$\sqrt{147} = \sqrt{49 \times 3}$$
$$=\sqrt{49} \times \sqrt{3}$$
$$=\sqrt{7 \times 7} \times \sqrt{3}$$
$$=7 \times \sqrt{3}$$
$$=7\sqrt{3}$$

Question 9: Simplify $\sqrt{600}$

To achieve this, find the factors of 600 , such that one is a perfect square.

$$\sqrt{600} = \sqrt{100 \times 6}$$
$$=\sqrt{100} \times \sqrt{6}$$
$$=\sqrt{10 \times 10} \times \sqrt{6}$$
$$=10 \times \sqrt{6}$$
$$=10\sqrt{6}$$

Question 10: Simplify $\sqrt{512}$

To achieve this, find the factors of 512 , such that one is a perfect square.

$$\sqrt{512} = \sqrt{256 \times 2}$$

$$=\sqrt{256} \times \sqrt{2}$$

$$=\sqrt{16 \times 16} \times \sqrt{2}$$

$$=16 \times \sqrt{2}$$

$$\mathbf{=16\sqrt{2}}$$

CONVERSION OF SURDS IN IT LOWEST FORM TO A SINGLE SURD

For conversion into a single surd

   a. *Express the rational coefficient as a square.*
   b. *Find the square root of the perfect square.*

Questions: *Express the following as a square root of a single surd.*

   a. $2\sqrt{5}$
   b. $7\sqrt{3}$
   c. $5\sqrt{7}$
   d. $3\sqrt{10}$
   e. $6\sqrt{6}$

Solution

   a. $2\sqrt{5}$

Express the following as a square root of single number.

$$2\sqrt{5} = \sqrt{2 \times 2} \times \sqrt{5}$$

$$= \sqrt{4} \times \sqrt{5}$$

$$= \sqrt{4 \times 5}$$

$$= \sqrt{20}$$

Therefore;

$$2\sqrt{5} = \sqrt{20}$$

b. $7\sqrt{3}$

Express the following as a square root of single number.

$$7\sqrt{3} = \sqrt{7 \times 7} \times \sqrt{3}$$

$$= \sqrt{49} \times \sqrt{3}$$

$$= \sqrt{49 \times 3}$$

$$= \sqrt{147}$$

Therefore;

$$7\sqrt{3} = \sqrt{147}$$

c. $5\sqrt{7}$

Express the following as a square root of single number.

$$5\sqrt{7} = \sqrt{5 \times 5} \times \sqrt{7}$$

$$= \sqrt{25} \times \sqrt{7}$$

$$= \sqrt{25 \times 7}$$

$$= \sqrt{175}$$

Therefore;

$$5\sqrt{7} = \sqrt{175}$$

d. $3\sqrt{10}$

Express the following as a square root of single number.

$$3\sqrt{10} = \sqrt{3 \times 3} \times \sqrt{10}$$
$$= \sqrt{9} \times \sqrt{10}$$
$$= \sqrt{9 \times 10}$$
$$= \sqrt{90}$$

Therefore;

$$\mathbf{3\sqrt{10} = \sqrt{90}}$$

e. $6\sqrt{6}$

Express the following as a square root of single number.

$$6\sqrt{6} = \sqrt{6 \times 6} \times \sqrt{6}$$
$$= \sqrt{36} \times \sqrt{6}$$
$$= \sqrt{36 \times 6}$$
$$= \sqrt{216}$$

Therefore;

$$\mathbf{6\sqrt{6} = \sqrt{216}}$$

Exercise 4.1

Simplify the following.

1. $\sqrt{48}$

2. $\sqrt{20}$

3. $\sqrt{50}$

4. $\sqrt{99}$

5. $\sqrt{150}$

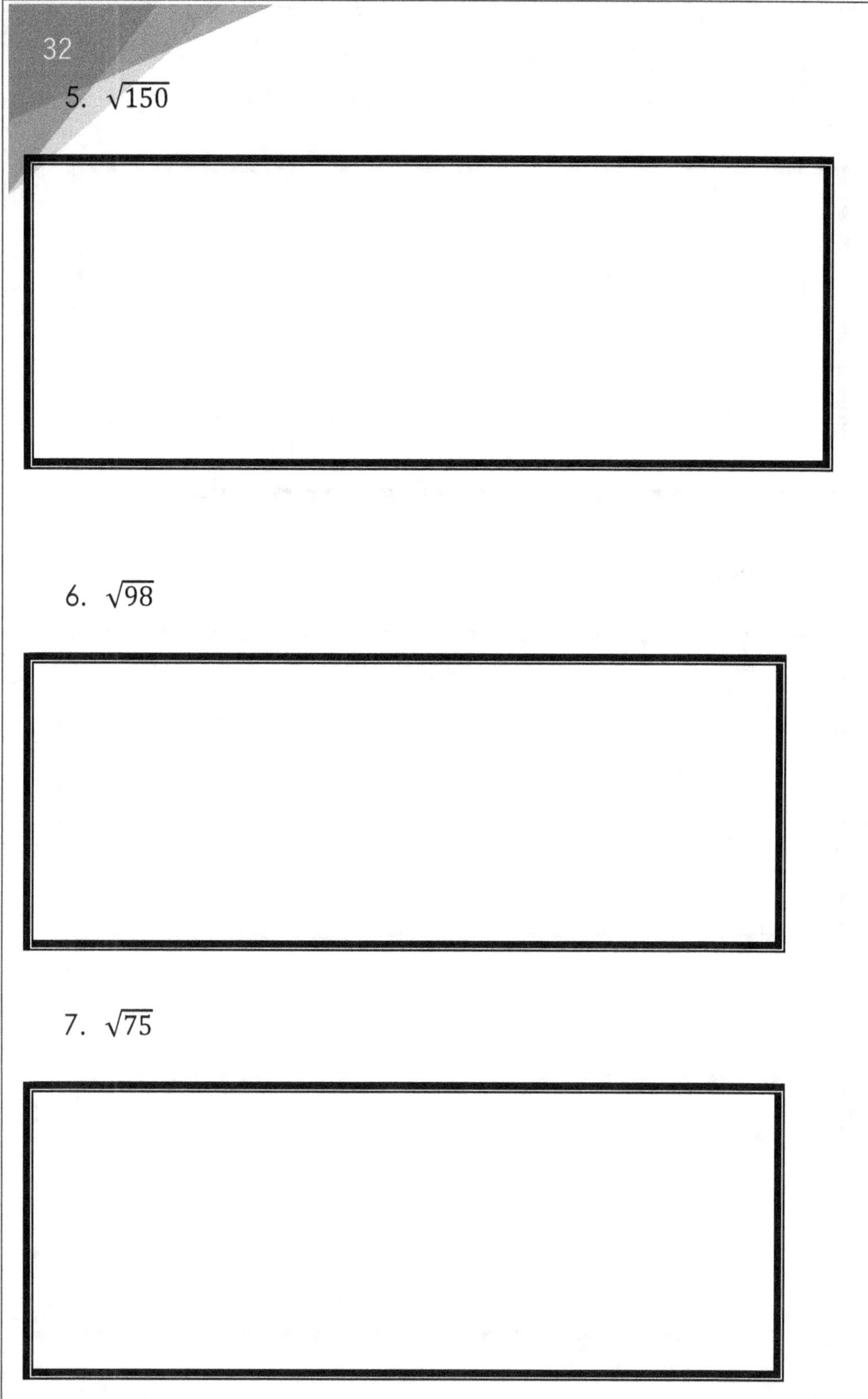

6. $\sqrt{98}$

7. $\sqrt{75}$

8. $\sqrt{648}$

9. $\sqrt{800}$

10. $\sqrt{243}$

Exercise 4.2

Express the following as a squareroot of a single surd

1. $8\sqrt{7}$

2. $5\sqrt{8}$

3. $11\sqrt{3}$

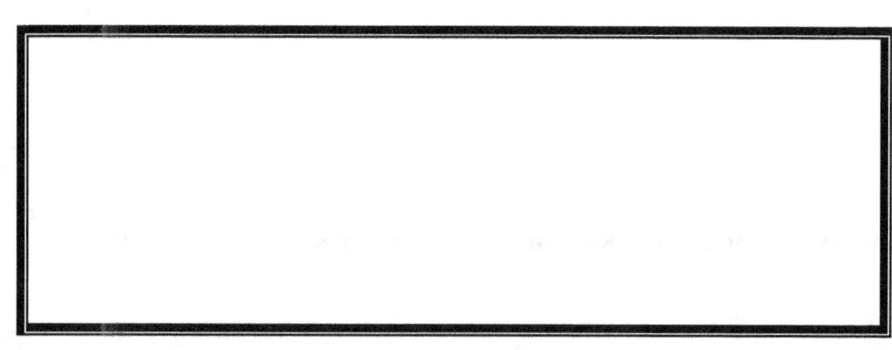

4. $7\sqrt{5}$

5. $9\sqrt{11}$

6. $10\sqrt{2}$

7. $12\sqrt{6}$

8. $9\sqrt{10}$

9. $5\sqrt{2}$

10. $12\sqrt{3}$

Chapter 5

OPERATIONS WITH SURDS

In this chapter, we shall discuss how you can add, subtract, multiply and divide surds.

ADDITION AND SUBTRACTION OF SURDS

Two or more surds can be added together or subtracted from one another if they are like surds, but the operation of addition or subtraction of unlike surds will still be equal to the same surds. Note that before the sum of two or more surds or the difference of surds, it must be simplified first to it basic form as explained in the previous chapter.

Question 1: Simplify $7\sqrt{2} + 3\sqrt{2}$

$$7\sqrt{2} + 3\sqrt{2} = (7 + 3)\sqrt{2}$$
$$= 10\sqrt{2}$$

Question 2: Simplify $5\sqrt{3} - 9\sqrt{3}$

$$5\sqrt{3} - 9\sqrt{3} = (5 - 9)\sqrt{3}$$
$$= -4\sqrt{3}$$

Question 3: Simplify $\sqrt{12} - \sqrt{3}$

$$\sqrt{12} - \sqrt{3}$$

Simplify surds

$$\sqrt{12} = \sqrt{4 \times 3}$$
$$=\sqrt{4} \times \sqrt{3}$$
$$=\sqrt{2 \times 2} \times \sqrt{3}$$
$$=2 \times \sqrt{3}$$
$$=\mathbf{2\sqrt{3}}$$

Hence;

$$\sqrt{12} - \sqrt{3} = 2\sqrt{3} - \sqrt{3}$$
$$(2 - 1)\sqrt{3} = \sqrt{3}$$

Question 4: Simplify $\sqrt{175} - 4\sqrt{7}$

$$\sqrt{175} - 4\sqrt{7}$$

Simplify surds

$$\sqrt{175} = \sqrt{25 \times 7}$$
$$=\sqrt{25} \times \sqrt{7}$$
$$=\sqrt{5 \times 5} \times \sqrt{7}$$
$$=5 \times \sqrt{7}$$
$$=\mathbf{5\sqrt{7}}$$

Hence;

$$\sqrt{175} - 4\sqrt{7} = 5\sqrt{7} - 4\sqrt{7}$$
$$(5 - 4)\sqrt{7} = \sqrt{7}$$

Question 5: Simplify $\sqrt{99} - \sqrt{44} - \sqrt{11}$

$$\sqrt{99} - \sqrt{44} - \sqrt{11}$$

Simplify surds

$$\sqrt{99} = \sqrt{9 \times 11}$$
$$=\sqrt{9} \times \sqrt{11}$$
$$=\sqrt{3 \times 3} \times \sqrt{11}$$
$$=3 \times \sqrt{11}$$
$$=\mathbf{3\sqrt{11}}$$

$$\sqrt{44} = \sqrt{4 \times 11}$$
$$=\sqrt{4} \times \sqrt{11}$$
$$=\sqrt{2 \times 2} \times \sqrt{11}$$
$$=2 \times \sqrt{11}$$
$$=\mathbf{2\sqrt{11}}$$

Hence;

$$\sqrt{99} - \sqrt{44} - \sqrt{11} = 3\sqrt{11} - 2\sqrt{11} - \sqrt{11}$$
$$(3 - 2 - 1)\sqrt{7} = \mathbf{0\sqrt{7} = 0}$$

Question 6: Simplify $2\sqrt{54} + \sqrt{24} - \sqrt{216}$

$$2\sqrt{54} + \sqrt{24} - \sqrt{216}$$

Simplify surds

$$\mathbf{2\sqrt{54}} = 2\sqrt{9 \times 6}$$
$$=2\sqrt{9} \times \sqrt{6}$$
$$=2\sqrt{3 \times 3} \times \sqrt{6}$$
$$=2 \times 3 \times \sqrt{6}$$
$$=\mathbf{6\sqrt{6}}$$

$$\sqrt{24} = \sqrt{4 \times 6}$$
$$=\sqrt{4} \times \sqrt{6}$$
$$=\sqrt{2 \times 2} \times \sqrt{6}$$
$$=2 \times \sqrt{6}$$
$$=\mathbf{2\sqrt{6}}$$

$$\sqrt{216} = \sqrt{36 \times 6}$$
$$=\sqrt{36} \times \sqrt{6}$$
$$=\sqrt{6 \times 6} \times \sqrt{6}$$
$$=6 \times \sqrt{6}$$
$$=\mathbf{6\sqrt{6}}$$

Hence;

$$2\sqrt{54} + \sqrt{24} - \sqrt{216} = 6\sqrt{6} + 2\sqrt{6} - 6\sqrt{6}$$
$$(6 + 2 - 6)\mathbf{\sqrt{6}} = \mathbf{2\sqrt{6}}$$

Question 7: Simplify $2\sqrt{135} - 2\sqrt{60} + \sqrt{15} - \sqrt{240}$

$$2\sqrt{135} - 2\sqrt{60} + \sqrt{15} - \sqrt{240}$$

Simplify surds

$$\mathbf{2\sqrt{135}} = 2\sqrt{9 \times 15}$$
$$=2\sqrt{9} \times \sqrt{15}$$
$$=2\sqrt{3 \times 3} \times \sqrt{15}$$
$$=2 \times 3 \times \sqrt{15}$$

$$=6\sqrt{15}$$

$$2\sqrt{60} = 2\sqrt{4 \times 15}$$
$$=2 \times \sqrt{4} \times \sqrt{15}$$
$$=2 \times \sqrt{2 \times 2} \times \sqrt{15}$$
$$=2 \times 2 \times \sqrt{15}$$
$$=4\sqrt{15}$$

$$\sqrt{240} = \sqrt{16 \times 15}$$
$$=\sqrt{16} \times \sqrt{15}$$
$$=\sqrt{4 \times 4} \times \sqrt{15}$$
$$=4 \times \sqrt{15}$$
$$=4\sqrt{15}$$

Hence;

$$2\sqrt{135} - 2\sqrt{60} + \sqrt{15} - \sqrt{240} = 6\sqrt{15} - 4\sqrt{15} + \sqrt{15} - 4\sqrt{15}$$

$$(6 - 4 + 1 - 4)\sqrt{15} = -1\sqrt{15} = -\sqrt{15}$$

Question 8: Simplify $2\sqrt{8} - 3\sqrt{32} + 4\sqrt{50}$

$$2\sqrt{8} - 3\sqrt{32} + 4\sqrt{50}$$

Simplify surds

$$2\sqrt{8} = 2\sqrt{4 \times 2}$$
$$=2\sqrt{4} \times \sqrt{2}$$

$$=2\sqrt{2 \times 2} \times \sqrt{2}$$

$$=2 \times 2 \times \sqrt{2}$$

$$=\mathbf{4\sqrt{2}}$$

$$\mathbf{3\sqrt{32}} = 3\sqrt{16 \times 2}$$

$$=3 \times \sqrt{16} \times \sqrt{2}$$

$$=3 \times \sqrt{4 \times 4} \times \sqrt{2}$$

$$=3 \times 4 \times \sqrt{2}$$

$$=\mathbf{12\sqrt{2}}$$

$$\mathbf{4\sqrt{50}} = 4\sqrt{25 \times 2}$$

$$=4\sqrt{25} \times \sqrt{2}$$

$$=4\sqrt{5 \times 5} \times \sqrt{2}$$

$$=4 \times 5 \times \sqrt{2}$$

$$=\mathbf{20\sqrt{2}}$$

Hence;

$$2\sqrt{8} - 3\sqrt{32} + 4\sqrt{50} = 4\sqrt{2} - 12\sqrt{2} + 20\sqrt{2}$$

$$(4 - 12 + 20)\sqrt{\mathbf{2}} = \mathbf{12\sqrt{2}}$$

Question 8: Simplify $2\sqrt{8} - 3\sqrt{32} + 4\sqrt{50}$

$$2\sqrt{8} - 3\sqrt{32} + 4\sqrt{50}$$

Simplify surds

$$2\sqrt{8} = 2\sqrt{4 \times 2}$$
$$= 2\sqrt{4} \times \sqrt{2}$$
$$= 2\sqrt{2 \times 2} \times \sqrt{2}$$
$$= 2 \times 2 \times \sqrt{2}$$
$$= 4\sqrt{2}$$

$$3\sqrt{32} = 3\sqrt{16 \times 2}$$
$$= 3 \times \sqrt{16} \times \sqrt{2}$$
$$= 3 \times \sqrt{4 \times 4} \times \sqrt{2}$$
$$= 3 \times 4 \times \sqrt{2}$$
$$= 12\sqrt{2}$$

$$4\sqrt{50} = 4\sqrt{25 \times 2}$$
$$= 4\sqrt{25} \times \sqrt{2}$$
$$= 4\sqrt{5 \times 5} \times \sqrt{2}$$
$$= 4 \times 5 \times \sqrt{2}$$
$$= 20\sqrt{2}$$

Hence;

$$2\sqrt{8} - 3\sqrt{32} + 4\sqrt{50} = 4\sqrt{2} - 12\sqrt{2} + 20\sqrt{2}$$
$$(4 - 12 + 20)\sqrt{2} = 12\sqrt{2}$$

MULIPLICATION OF SURDS

To multiply two or more surds, they must first be simplified. After simplification, multiply whole number with whole numbers and multiply surd with surds.

SINGLE SURD

Question 1: Simplify $\sqrt{27} \times \sqrt{72}$

$\sqrt{27} \times \sqrt{72}$

$$\sqrt{27} \times \sqrt{72} = \sqrt{9 \times 3} \times \sqrt{36 \times 2}$$

$$\sqrt{9} \times \sqrt{3} \times \sqrt{36} \times \sqrt{2}$$

$$\sqrt{3 \times 3} \times \sqrt{3} \times \sqrt{6 \times 6} \times \sqrt{2}$$

$$3\sqrt{3} \times 6\sqrt{2}$$

$$3 \times 6 \times \sqrt{3} \times \sqrt{2}$$

$$18 \times \sqrt{3 \times 2}$$

$$18 \times \sqrt{6} = \mathbf{18\sqrt{6}}$$

Question 2: Simplify $\left(3\sqrt{7}\right)^2$

$\left(3\sqrt{7}\right)^2$

$$\left(3\sqrt{7}\right)^2 = 3\sqrt{7} \times 3\sqrt{7}$$

$$3 \times 3 \times \sqrt{7} \times \sqrt{7}$$

$$9 \times \sqrt{7 \times 7}$$

$$9 \times 7 = \mathbf{63}$$

Question 3: Simplify $\sqrt{10} \times 5\sqrt{2} \times \sqrt{20}$

$$\sqrt{10} \times 5\sqrt{2} \times \sqrt{20}$$

$$\mathbf{\sqrt{10} \times 5\sqrt{2}} \times \sqrt{20} = \sqrt{5 \times 2} \times 5\sqrt{2} \times \sqrt{5 \times 4}$$

$$\sqrt{5} \times \sqrt{2} \times 5 \times \sqrt{2} \times \sqrt{5} \times \sqrt{4}$$

$$5 \times \sqrt{5} \times \sqrt{5} \times 2 \times \sqrt{2} \times \sqrt{2}$$

$$5 \times 5 \times 2 \times 2 = \mathbf{100}$$

$$\sqrt{10} \times 5\sqrt{2} \times \sqrt{20} = \mathbf{100}$$

$$3 \times 6 \times \sqrt{3} \times \sqrt{2}$$

$$18 \times \sqrt{3 \times 2}$$

$$18 \times \sqrt{6} = \mathbf{18\sqrt{6}}$$

Question 4: Simplify $\sqrt{5} \times \sqrt{3} \times \sqrt{12}$

$$\sqrt{5} \times \sqrt{3} \times \sqrt{12}$$

$$\mathbf{\sqrt{5} \times \sqrt{3} \times \sqrt{12}} = \sqrt{5} \times \sqrt{3} \times \sqrt{3 \times 4}$$

$$= \sqrt{5} \times \sqrt{3} \times \sqrt{3} \times \sqrt{4}$$

$$= \sqrt{5} \times \sqrt{3} \times \sqrt{3} \times 2$$

$$= \sqrt{5} \times 3 \times 2$$

$$= \sqrt{5} \times 6$$

$$= 6\sqrt{5}$$

$$\sqrt{5} \times \sqrt{3} \times \sqrt{12} = 6\sqrt{5}$$

Question 5: Simplify $\dfrac{3\sqrt{5} \times 5\sqrt{2}}{\sqrt{10}}$

$$\frac{3\sqrt{5} \times 5\sqrt{2}}{\sqrt{10}} = \frac{3 \times 5 \times \sqrt{5} \times \sqrt{2}}{\sqrt{10}}$$

$$= \frac{15 \times \sqrt{5 \times 2}}{\sqrt{10}}$$

$$= \frac{15 \times \sqrt{10}}{\sqrt{10}} = 15$$

MULTIPLICATION OF SURDS INVOLVING BRACKET

Multiplying surds can be achieved using algebraic method.

$$(a + b)(c + d) = (a \times c) + (a \times d) + (b \times c) + (b \times d)$$

Question 1: Simplify $(3\sqrt{5} + 2)(\sqrt{5} + 3)$

$(3\sqrt{5} + 2)(\sqrt{5} + 3)$

$(3\sqrt{5} + 2)(\sqrt{5} + 3) = (3\sqrt{5} \times \sqrt{5}) + (3\sqrt{5} \times 3) + 2 \times \sqrt{5} + 2 \times 3$

$\qquad (3\sqrt{25}) + 9\sqrt{5} + 2\sqrt{5} + 6$

$\qquad 3 \times 5 + 9\sqrt{5} + 2\sqrt{5} + 6$

$\qquad 15 + 11\sqrt{5} + 6$

$\qquad 15 + 6 + 11\sqrt{5}$

$\qquad = 21 + 11\sqrt{5}$

Question 2: Simplify $\left(3\sqrt{5} - \sqrt{5}\right)^2$

$$\left(3\sqrt{5} - \sqrt{5}\right)^2 = \left(3\sqrt{5} - \sqrt{5}\right)\left(3\sqrt{5} - \sqrt{5}\right)$$

$(3\sqrt{5} - \sqrt{5})(3\sqrt{5} - \sqrt{5}) = (3\sqrt{5} \times 3\sqrt{5}) - (3\sqrt{5} \times \sqrt{5}) - (\sqrt{5} \times$

$\qquad 3\sqrt{5}) + (\sqrt{5} \times \sqrt{5})$

$$(9 \times 3 \times \sqrt{5 \times 5}) - (3\sqrt{5 \times 5}) - (3\sqrt{5 \times 5}) + (\sqrt{5 \times 5})$$

$$=(27 \times 5) - (3 \times 5) - (3 \times 5) + (5)$$

$$= (135) - (15) - (15) + (5) = \mathbf{110}$$

$$\left(3\sqrt{5} - \sqrt{5}\right)^2 = \mathbf{110}$$

Question 3: Simplify $\left(2\sqrt{2} + \sqrt{5}\right)\left(\sqrt{5} - 2\sqrt{2}\right)$

$$\left(2\sqrt{2} + \sqrt{5}\right)\left(\sqrt{5} - 2\sqrt{2}\right)$$

$$=(2\sqrt{2} \times \sqrt{5}) - (2\sqrt{2} \times 2\sqrt{2}) + (\sqrt{5} \times \sqrt{5}) - (\sqrt{5} \times 2\sqrt{2})$$

$$(2 \times \sqrt{2 \times 5}) - (2 \times 2 \times \sqrt{2 \times 2}) + (\sqrt{5 \times 5}) - (\sqrt{5} \times 2\sqrt{2})$$

$$=(2 \times \sqrt{10}) - (4 \times 2) + (5) - (2\sqrt{10})$$

$$= \left(2\sqrt{10}\right) - (8) + (5) - \left(2\sqrt{10}\right)$$

$$-8 + 5 = \mathbf{-3}$$

Question 4: Simplify $\left(\sqrt{6} - \sqrt{5}\right)\left(\sqrt{8} + \sqrt{15}\right)$

$$\left(\sqrt{6} - \sqrt{5}\right)\left(\sqrt{8} + \sqrt{15}\right)$$

$$=(\sqrt{6} \times \sqrt{8}) + (\sqrt{6} \times \sqrt{15}) - (\sqrt{5} \times \sqrt{8}) - (\sqrt{5} \times \sqrt{15})$$

$$\left(\sqrt{48}\right) + \left(\sqrt{90}\right) - \left(\sqrt{40}\right) - \left(\sqrt{75}\right)$$

$$=\left(\sqrt{16 \times 3}\right) + \left(\sqrt{9 \times 10}\right) - \left(\sqrt{4 \times 10}\right) - \left(\sqrt{25 \times 3}\right)$$

$$= \left(\sqrt{16} \times \sqrt{3}\right) + \left(\sqrt{9} \times \sqrt{10}\right) - \left(\sqrt{4} \times \sqrt{10}\right) - \left(\sqrt{25} \times \sqrt{3}\right)$$

$$= \left(4\sqrt{3}\right) + \left(3\sqrt{10}\right) - \left(2\sqrt{10}\right) - \left(5\sqrt{3}\right)$$

$$= \sqrt{\mathbf{10}} - \sqrt{\mathbf{3}}$$

$$\left(\sqrt{6} - \sqrt{5}\right)\left(\sqrt{8} + \sqrt{15}\right) = \sqrt{\mathbf{10}} - \sqrt{\mathbf{3}}$$

Question 5: Simplify $\left(\sqrt{2} + \frac{1}{\sqrt{2}}\right)^2$

$$\left(\sqrt{2} + \frac{1}{\sqrt{2}}\right)^2$$

$$\left(\sqrt{2} + \frac{1}{\sqrt{2}}\right)\left(\sqrt{2} + \frac{1}{\sqrt{2}}\right)$$

$$= \left(\sqrt{2} \times \sqrt{2}\right) + \left(\sqrt{2} \times \frac{1}{\sqrt{2}}\right) + \left(\sqrt{2} \times \frac{1}{\sqrt{2}}\right) + \left(\frac{1}{\sqrt{2}} \times \frac{1}{\sqrt{2}}\right)$$

$$(2) + (1) + (1) - \left(\frac{1}{\sqrt{4}}\right)$$

$$= 2 + 1 + 1 + \frac{1}{2}$$

$$= 4\frac{1}{2} = 4.5$$

$$\left(\sqrt{2} + \frac{1}{\sqrt{2}}\right)^2 = 4\frac{1}{2} = 4.5$$

Exercise 5.1

Simplify the following.

1. $2\sqrt{12} + 3\sqrt{48} - \sqrt{75}$

2. $4\sqrt{8} - 3\sqrt{98} + \sqrt{128}$

3. $3\sqrt{18} + 4\sqrt{72}$

4. $\sqrt{175} - 5\sqrt{7} + 4\sqrt{7}$

5. $\sqrt{52} - \sqrt{117}$

6. $\sqrt{224} - \sqrt{126} - \sqrt{56}$

7. $40\sqrt{2} + \sqrt{8}$

Exercise 5.2

Given that $\sqrt{2} = 1.414$, $\sqrt{3} = 1.732$, $\sqrt{5} = 2.236$. *Evaluate each of the following, leave your answer in 4 significant figure.*

1. $\sqrt{18}$

2. $\sqrt{40}$

3. $3\sqrt{3} + \sqrt{75}$

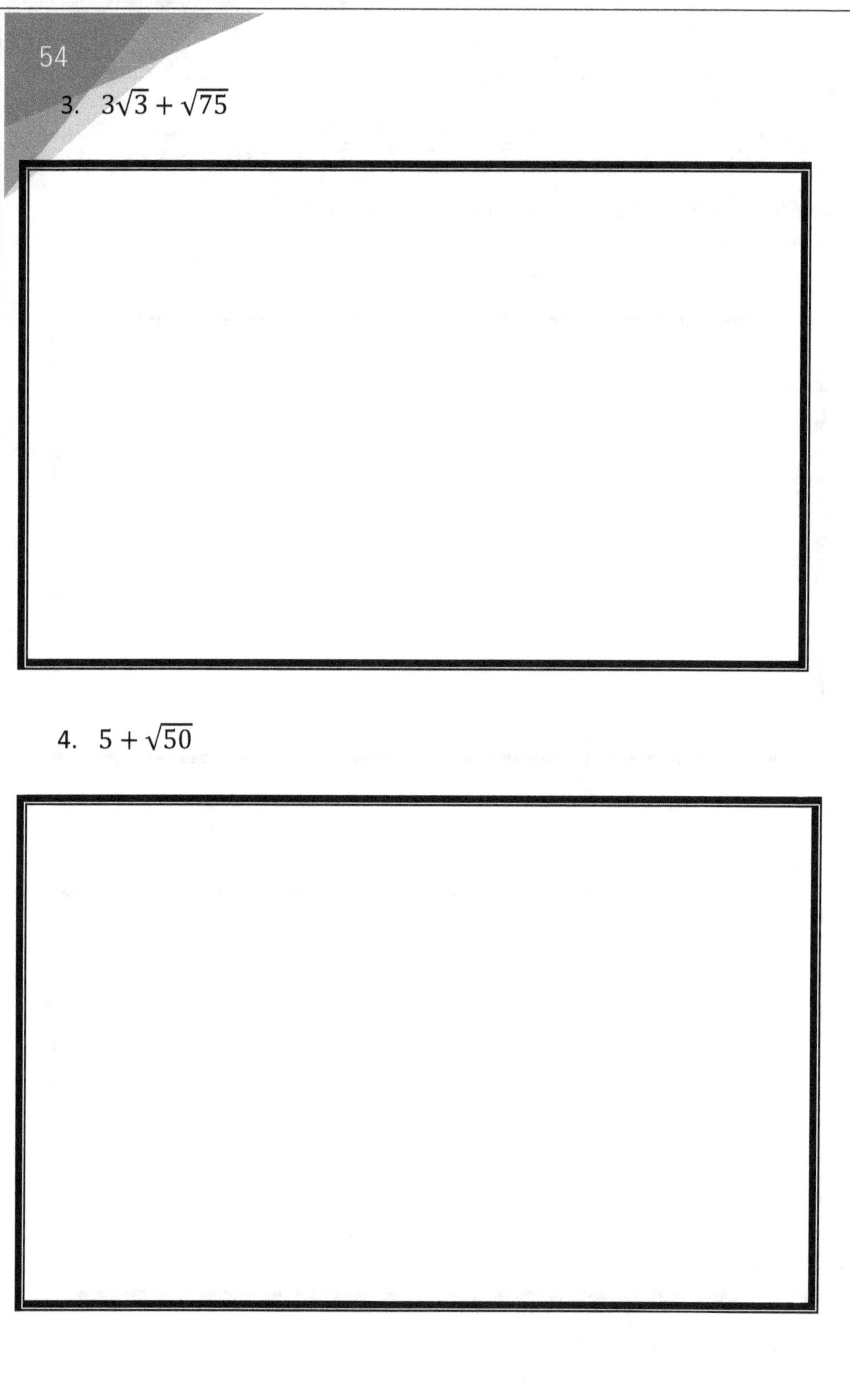

4. $5 + \sqrt{50}$

5. $\sqrt{125}$

Exercise 5.3

Simplify the following.

1. $\left(2\sqrt{6} - 1\right)\left(\sqrt{3} - \sqrt{2}\right)$

2. $\left(4\sqrt{3} + 2\sqrt{2}\right)\left(\mathbf{3}\sqrt{2} + 2\sqrt{5}\right)$

3. $2\sqrt{5}\left(\mathbf{3}\sqrt{5} - 2\sqrt{2}\right)$

4. $\left(\sqrt{7} - \sqrt{3}\right)\left(2\sqrt{7} + \sqrt{3}\right)$

5. $\left(\sqrt{5} - 2\sqrt{10}\right)^2$

6. $\sqrt{6} \times \sqrt{8} \times \sqrt{10} \times \sqrt{12}$

7. $\left(2\sqrt{11}\right)^2$

8. $\sqrt{5} \times \sqrt{24} \times \sqrt{30}$

9. $\sqrt{15} \times \sqrt{12}$

10. $\sqrt{6}(\sqrt{2} + \sqrt{6})$

Chapter 6

CONJUGATE SURDS

This can also be referred to as conjugate of a **binomial surd**. An expression of surd is binomial when one or two of the term contain surds.

Example: $7 - \sqrt{5}$, $3\sqrt{2} + \sqrt{5}$.

To rationalize a binomial surd, we use difference of two square.

Example

To rationalize $\sqrt{a} + b$, we multiply it by it conjugate - $(\sqrt{a} - b)$, therefore making $(\sqrt{a} + b)(\sqrt{a} - b)$ the conjugate of each other

Two surds are said to be conjugate is their product give rise to a rational number.

Example:

The conjugate of

a) $\sqrt{3} - \sqrt{2}$ *is* $\sqrt{3} + \sqrt{2}$

b) $2\sqrt{5} + 3\sqrt{3}$ *is* $2\sqrt{2} - 3\sqrt{3}$

Examples : Multiply the following binomial surd with it conjugate

   a. $\sqrt{3} - \sqrt{2}$

   b. $\sqrt{5} - 3\sqrt{2}$

   c. $\sqrt{2} + 7$

Solution

a. $\sqrt{3} - \sqrt{2}$

The conjugate of $\sqrt{3} - \sqrt{2}$ is $\sqrt{3} + \sqrt{2}$

$$(\sqrt{3} - \sqrt{2})(\sqrt{3} + \sqrt{2})$$

From the analysis below

$$(a + b)(c + d) = (a \times c) + (a \times d) + (b \times c) + (b \times d)$$

$$(\sqrt{3} - \sqrt{2})(\sqrt{3} + \sqrt{2})$$

$$= (\sqrt{3} \times \sqrt{3}) + (\sqrt{3} \times \sqrt{2}) - (\sqrt{2} \times \sqrt{3})$$

$$- (\sqrt{2} \times \sqrt{2})$$

$$(\sqrt{3 \times 3}) + (\sqrt{3 \times 2}) - (\sqrt{2 \times 3}) - (\sqrt{2 \times 2})$$

$$(\sqrt{9}) + (\sqrt{6}) - (\sqrt{6}) - (\sqrt{4})$$

$$= (3) + (\sqrt{6}) - (\sqrt{6}) - (2)$$

$$= (3) - (2) = 1$$

b. $\sqrt{5} - 3\sqrt{2}$

The conjugate of $\sqrt{5} - 3\sqrt{2}$ is $\sqrt{5} + 3\sqrt{2}$

$$(\sqrt{5} - 3\sqrt{2})(\sqrt{5} + 3\sqrt{2})$$

From the analysis below

$$(a + b)(c + d) = (a \times c) + (a \times d) + (b \times c) + (b \times d)$$

$$(\sqrt{5} - 3\sqrt{2})(\sqrt{5} + 3\sqrt{2})$$

$$= (\sqrt{5} \times \sqrt{5}) + (\sqrt{5} \times 3\sqrt{2}) - (3\sqrt{2} \times \sqrt{5})$$

$$- (3\sqrt{2} \times 3\sqrt{2})$$

$$= (\sqrt{5 \times 5}) + (3\sqrt{5 \times 2}) - (3\sqrt{2 \times 5}) - (3 \times 3\sqrt{2 \times 2})$$

$$= \left(\sqrt{25}\right) + \left(3\sqrt{10}\right) - \left(3\sqrt{10}\right) - \left(9\sqrt{4}\right)$$

$$= (5) + \left(3\sqrt{10}\right) - \left(3\sqrt{10}\right) - (9 \times 2)$$

$$= (5) - (18) = -13$$

c. $\sqrt{2} + \sqrt{7}$

The conjugate of $\sqrt{2} + \sqrt{7}$ is $\sqrt{2} - \sqrt{7}$

$$\left(\sqrt{2} + \sqrt{7}\right)\left(\sqrt{2} - \sqrt{7}\right)$$

From the analysis below

$$(a + b)(c + d) = (a \times c) + (a \times d) + (b \times c) + (b \times d)$$

$$\left(\sqrt{2} + \sqrt{7}\right)\left(\sqrt{2} - \sqrt{7}\right)$$

$$= \left(\sqrt{2} \times \sqrt{2}\right) - \left(\sqrt{2} \times \sqrt{7}\right) + \left(\sqrt{7} \times \sqrt{2}\right)$$

$$- \left(\sqrt{7} \times \sqrt{7}\right)$$

$$= \left(\sqrt{2 \times 2}\right) - \left(\sqrt{2 \times 7}\right) - \left(\sqrt{7 \times 2}\right) - \left(\sqrt{7 \times 7}\right)$$

$$= \left(\sqrt{4}\right) - \left(\sqrt{14}\right) - \left(\sqrt{14}\right) - \left(\sqrt{49}\right)$$

$$= (2) - \left(\sqrt{14}\right) + \left(\sqrt{14}\right) - (7)$$

$$= (5) - (7) = -5$$

RATIONALISING THE DENOMINATOR

When you are to rationalize, it means you are to remove any radical expressions in the denominator. To achieve this, the numerator and the denominator of the major fraction is multiplied by the equivalent radical expression of the denominator, thereby removing the radical expression in the denominator.

Steps to rationalize

1. *Multiply the numerator and the denominator of the original fraction with the radical expression in the denominator.*

2. *Simplify all radicals.*

3. *Simplify the fraction if necessary.*

Rationalizing the denominator with a single surd

Question 1 : Rationalise $\dfrac{5}{\sqrt{2}}$

To rationalized, *Multiply the numerator and the denominator of the original fraction with the radical expression in the denominator, then simplify.*

$$\frac{5}{\sqrt{2}} \times \frac{\sqrt{2}}{\sqrt{2}}$$

$$= \frac{5\sqrt{2}}{\sqrt{2} \times \sqrt{2}}$$

$$= \frac{5\sqrt{2}}{\sqrt{4}}$$

$$= \frac{5\sqrt{2}}{2}$$

$$\frac{5}{\sqrt{2}} = \frac{5\sqrt{2}}{2}$$

Question 2 : Rationalise $\dfrac{7}{\sqrt{3}}$

To rationalized, *Multiply the numerator and the denominator of the original fraction with the radical expression in the denominator, then simplify.*

$$\frac{7}{\sqrt{3}} \times \frac{\sqrt{3}}{\sqrt{3}}$$

$$= \frac{7\sqrt{3}}{\sqrt{3} \times \sqrt{3}}$$

$$= \frac{7\sqrt{3}}{\sqrt{9}}$$

$$\frac{7}{\sqrt{3}} = \frac{7\sqrt{3}}{3}$$

$$\frac{5}{\sqrt{2}} = \frac{5\sqrt{2}}{2}$$

Question 3 : Rationalise $\dfrac{3+\sqrt{2}}{\sqrt{3}}$

To rationalized, *Multiply the numerator and the denominator of the original fraction with the radical expression in the denominator, then simplify.*

$$\frac{3 + \sqrt{2}}{\sqrt{3}}$$

$$\frac{3 + \sqrt{2}}{\sqrt{3}} \times \frac{\sqrt{3}}{\sqrt{3}} = \frac{\sqrt{3}\left(3 + \sqrt{2}\right)}{3}$$

$$= \frac{3\sqrt{3} + \sqrt{6}}{3}$$

$$\frac{3 + \sqrt{2}}{\sqrt{3}} = \frac{3\sqrt{3} + \sqrt{6}}{3}$$

Question 4 : Rationalise $\dfrac{8}{4\sqrt{5}}$

To rationalized, *Multiply the numerator and the denominator of the original fraction with the radical expression in the denominator, then simplify.*

$$= \frac{8}{4\sqrt{5}} \times \frac{\sqrt{5}}{\sqrt{5}}$$

$$= \frac{8\sqrt{5}}{4 \times 5}$$

$$= \frac{8\sqrt{5}}{20} = \frac{2\sqrt{5}}{5}$$

$$\frac{8}{4\sqrt{5}} = \frac{2\sqrt{5}}{5}$$

Question 5 : Rationalise $\dfrac{3\sqrt{2}+2\sqrt{3}}{4\sqrt{2}}$

To rationalized, *Multiply the numerator and the denominator of the original fraction with the radical expression in the denominator, then simplify.*

$$\frac{3\sqrt{2} + 2\sqrt{3}}{4\sqrt{2}}$$

$$\frac{3\sqrt{2} + 2\sqrt{3}}{4\sqrt{2}} \times \frac{\sqrt{2}}{\sqrt{2}} = \frac{\sqrt{2}\left(3\sqrt{2} + 2\sqrt{3}\right)}{4 \times 2}$$

$$= \frac{\left(6 + 2\sqrt{6}\right)}{8}$$

$$= \frac{2(3+\sqrt{6})}{8} = 3 + \sqrt{6}$$

Question 6 : Express $\frac{\sqrt{2}+5}{\sqrt{10}}$ in the form $a\sqrt{5} + b\sqrt{10}$

$$\frac{\sqrt{2} + 5}{\sqrt{10}}$$

To rationalized, *Multiply the numerator and the denominator of the original fraction with the radical expression in the denominator, then simplify.*

$$\frac{\sqrt{2} + 5}{\sqrt{10}} \times \frac{\sqrt{10}}{\sqrt{10}} = \frac{\sqrt{10}(\sqrt{2} + 5)}{10}$$

$$= \frac{(\sqrt{20} + 5\sqrt{10})}{10}$$

$$= \frac{(\sqrt{4 \times 5} + 5\sqrt{10})}{10}$$

$$= \frac{\left((\sqrt{4} \times \sqrt{5}) + 5\sqrt{10}\right)}{10}$$

$$= \frac{\left((\sqrt{4} \times \sqrt{5}) + 5\sqrt{10}\right)}{10}$$

$$= \frac{\left((2 \times \sqrt{5}) + 5\sqrt{10}\right)}{10}$$

$$= \frac{\left((2\sqrt{5}) + 5\sqrt{10}\right)}{10}$$

$$= \frac{2\sqrt{5}}{10} + \frac{5\sqrt{10}}{10}$$

$$\frac{1}{5}\sqrt{5} + \frac{1}{2}\sqrt{10}$$

comparing $a\sqrt{5} + b\sqrt{10}$ *to the result*

$$a\sqrt{5} + b\sqrt{10} = \frac{1}{5}\sqrt{5} + \frac{1}{2}\sqrt{10}$$

where

$$a = \frac{1}{5}$$
$$b = \frac{1}{2}$$

Rationalizing the Denominator With Binomial Surds

To achieve this, you multiply both the numerator and the denominator by the conjugate of the denominator.

Question 1 : Rationalize $\frac{3+\sqrt{2}}{\sqrt{5}+\sqrt{3}}$

$$\frac{3 + \sqrt{2}}{\sqrt{5} + \sqrt{3}}$$

The conjugate of the denominator expression

$\sqrt{5} + \sqrt{3}$ is $\sqrt{5} - \sqrt{3}$

Multiply the conjugate of the denominator with the

numerator and the denominator $\sqrt{5} - \sqrt{3}$

$$\frac{3 + \sqrt{2}}{\sqrt{5} + \sqrt{3}} \times \frac{\sqrt{5} - \sqrt{3}}{\sqrt{5} - \sqrt{3}}$$

Use this analysis for the multiplication

$$(a + b)(c + d) = (a \times c) + (a \times d) + (b \times c) + (b \times d)$$

$$\frac{(3 \times \sqrt{5}) - (3 \times \sqrt{3}) + (\sqrt{2} \times \sqrt{5}) - (\sqrt{2} \times \sqrt{3})}{(\sqrt{5} \times \sqrt{5}) - (\sqrt{5} \times \sqrt{3}) + (\sqrt{3} \times \sqrt{5}) - (\sqrt{3} \times \sqrt{3})}$$

$$\frac{3\sqrt{5} - 3\sqrt{3} + \sqrt{10} - \sqrt{6}}{\sqrt{25} - \sqrt{15} + \sqrt{15} - \sqrt{9}}$$

$$\frac{3\sqrt{5} - 3\sqrt{3} + \sqrt{10} - \sqrt{6}}{5 - 3}$$

$$= \frac{3\sqrt{5} - 3\sqrt{3} + \sqrt{10} - \sqrt{6}}{2}$$

Question 2 : *By rationalizing the denominator, simplify*

$$\frac{5\sqrt{2}+3}{1-\sqrt{3}}$$

$$\frac{5\sqrt{2} + 3}{1 - \sqrt{3}}$$

The conjugate of the denominator expression

$$1 - \sqrt{3} \text{ is } 1 + \sqrt{3}$$

Multiply the conjugate of the denominator with the numerator and the denominator $1 + \sqrt{3}$

$$\frac{5\sqrt{2} + 3}{1 - \sqrt{3}} \times \frac{1 + \sqrt{3}}{1 + \sqrt{3}}$$

Use this analysis for the multiplication

$$(a + b)(c + d) = (a \times c) + (a \times d) + (b \times c) + (b \times d)$$

$$\frac{\left(5\sqrt{2} \times 1\right) + \left(5\sqrt{2} \times \sqrt{3}\right) + (3 \times 1) + \left(3 \times \sqrt{3}\right)}{(1 \times 1) + \left(1 \times \sqrt{3}\right) - \left(\sqrt{3} \times 1\right) - \left(\sqrt{3} \times \sqrt{3}\right)}$$

$$\frac{5\sqrt{2} + 5\sqrt{6} + 3 + 3\sqrt{3}}{1 - \sqrt{3} + \sqrt{3} - 3}$$

$$\frac{5\sqrt{2} + 5\sqrt{6} + 3 + 3\sqrt{3}}{1 - 3}$$

$$= \frac{5\sqrt{2} + 5\sqrt{6} + 3 + 3\sqrt{3}}{-2}$$

$$= \frac{-\left(5\sqrt{2} + 5\sqrt{6} + 3 + 3\sqrt{3}\right)}{2}$$

Question 3 : *By rationalizing the denominator, simplify*

$$\frac{2+4\sqrt{6}}{3\sqrt{2}-3}$$

$$\frac{2 + 4\sqrt{6}}{3\sqrt{2} - 3}$$

The conjugate of the denominator expression

$$3\sqrt{2} - 3 \text{ is } 3\sqrt{2} + 3$$

Multiply the conjugate of the denominator with the

numerator and the denominator $3\sqrt{2} + 3$

$$\frac{2 + 4\sqrt{6}}{3\sqrt{2} - 3} \times \frac{3\sqrt{2} + 3}{3\sqrt{2} + 3}$$

Use this analysis for the multiplication

$$(a + b)(c + d) = (a \times c) + (a \times d) + (b \times c) + (b \times d)$$

$$\frac{\left(2 \times 3\sqrt{2}\right) + (2 \times 3) + \left(4\sqrt{6} \times 3\sqrt{2}\right) + \left(4\sqrt{6} \times 3\right)}{\left(3\sqrt{2} \times 3\sqrt{2}\right) + \left(3\sqrt{2} \times 3\right) - \left(3 \times 3\sqrt{2}\right) - (3 \times 3)}$$

$$\frac{6\sqrt{2} + 6 + 12\sqrt{12} + 12\sqrt{6}}{9 \times 2 + 9\sqrt{2} - 9\sqrt{2} - 9}$$

$$\frac{6\sqrt{2} + 6 + \left(12 \times 2\sqrt{3}\right) + 12\sqrt{6}}{18 - 9}$$

$$= \frac{6\sqrt{2} + 6 + 24\sqrt{3} + 12\sqrt{6}}{9}$$

$$= \frac{6(\sqrt{2} + 1 + 4\sqrt{3} + 2\sqrt{6})}{9}$$

$$= \frac{2(\sqrt{2} + 1 + 4\sqrt{3} + 2\sqrt{6})}{3}$$

Question 4 : *By rationalizing the denominator, simplify*
$\frac{2}{3+\sqrt{7}}$

$$\frac{2}{3 + \sqrt{7}}$$

The conjugate of the denominator expression

$3 + \sqrt{7}$ **is** $3 - \sqrt{7}$

Multiply the conjugate of the denominator with the

numerator and the denominator $3 - \sqrt{7}$

$$\frac{2}{3 + \sqrt{7}} \times \frac{3 - \sqrt{7}}{3 - \sqrt{7}}$$

Use this analysis for the multiplication

$$(a + b)(c + d) = (a \times c) + (a \times d) + (b \times c) + (b \times d)$$

$$= \frac{(2 \times 3) - (2 \times \sqrt{7})}{(3 \times 3) - (3 \times \sqrt{7}) + (\sqrt{7} \times 3) - (\sqrt{7} \times \sqrt{7})}$$

$$= \frac{(6) - (2\sqrt{7})}{(9) - (3\sqrt{7}) + (3\sqrt{7}) - (7)}$$

$$= \frac{6 - 2\sqrt{7}}{(9) - (7)}$$

$$= \frac{2(3 - \sqrt{7})}{2}$$

$$= 3 - \sqrt{7}$$

$$\frac{2}{3+\sqrt{7}} = 3 - \sqrt{7}$$

Question 5 : *By rationalizing the denominator, simplify*

$\frac{7\sqrt{2}+3\sqrt{3}}{4\sqrt{2}-2\sqrt{3}}$

$$\frac{7\sqrt{2} + 3\sqrt{3}}{4\sqrt{2} - 2\sqrt{3}}$$

The conjugate of the denominator expression

$$4\sqrt{2} - 2\sqrt{3} \textbf{ is } 4\sqrt{2} + 2\sqrt{3}$$

Multiply the conjugate of the denominator with the

numerator and the denominator $4\sqrt{2} + 2\sqrt{3}$

$$\frac{7\sqrt{2} + 3\sqrt{3}}{4\sqrt{2} - 2\sqrt{3}} \times \frac{4\sqrt{2} + 2\sqrt{3}}{4\sqrt{2} + 2\sqrt{3}}$$

Use this analysis for the multiplication

$$(a + b)(c + d) = (a \times c) + (a \times d) + (b \times c) + (b \times d)$$

$$\frac{(7 \times 4 \times \sqrt{2} \times \sqrt{2}) + (7 \times 2 \times \sqrt{2} \times \sqrt{3}) + (3 \times 4 \times \sqrt{3} \times \sqrt{2}) + (3 \times 2 \times \sqrt{3} \times \sqrt{3})}{(4\sqrt{2} \times 4\sqrt{2}) + (4\sqrt{2} \times 2\sqrt{3}) - (2\sqrt{3} \times 4\sqrt{2}) - (2\sqrt{3} \times 2\sqrt{3})}$$

$$= \frac{(28 \times 2) + (14 \times \sqrt{6}) + (12 \times \sqrt{6}) + (6 \times 3)}{(16 \times 2) - (4 \times 3)}$$

$$= \frac{(56) + (14\sqrt{6}) + (12\sqrt{6}) + (18)}{(32) - (12)}$$

$$= \frac{56 + 18 + (14\sqrt{6}) + (12\sqrt{6})}{20}$$

$$\frac{74 + 26\sqrt{6}}{20} = \frac{74}{20} + \frac{26}{20}\sqrt{6}$$

$$= \frac{37}{10} + \frac{13}{10}\sqrt{6}$$

Exercise 6.1

Write and multiply the following binomial surd with it conjugate.

1. $\left(3\sqrt{5} + 4\right)$

2. $\left(2\sqrt{2} - 1\right)$

3. $\left(\sqrt{5} - \sqrt{3}\right)$

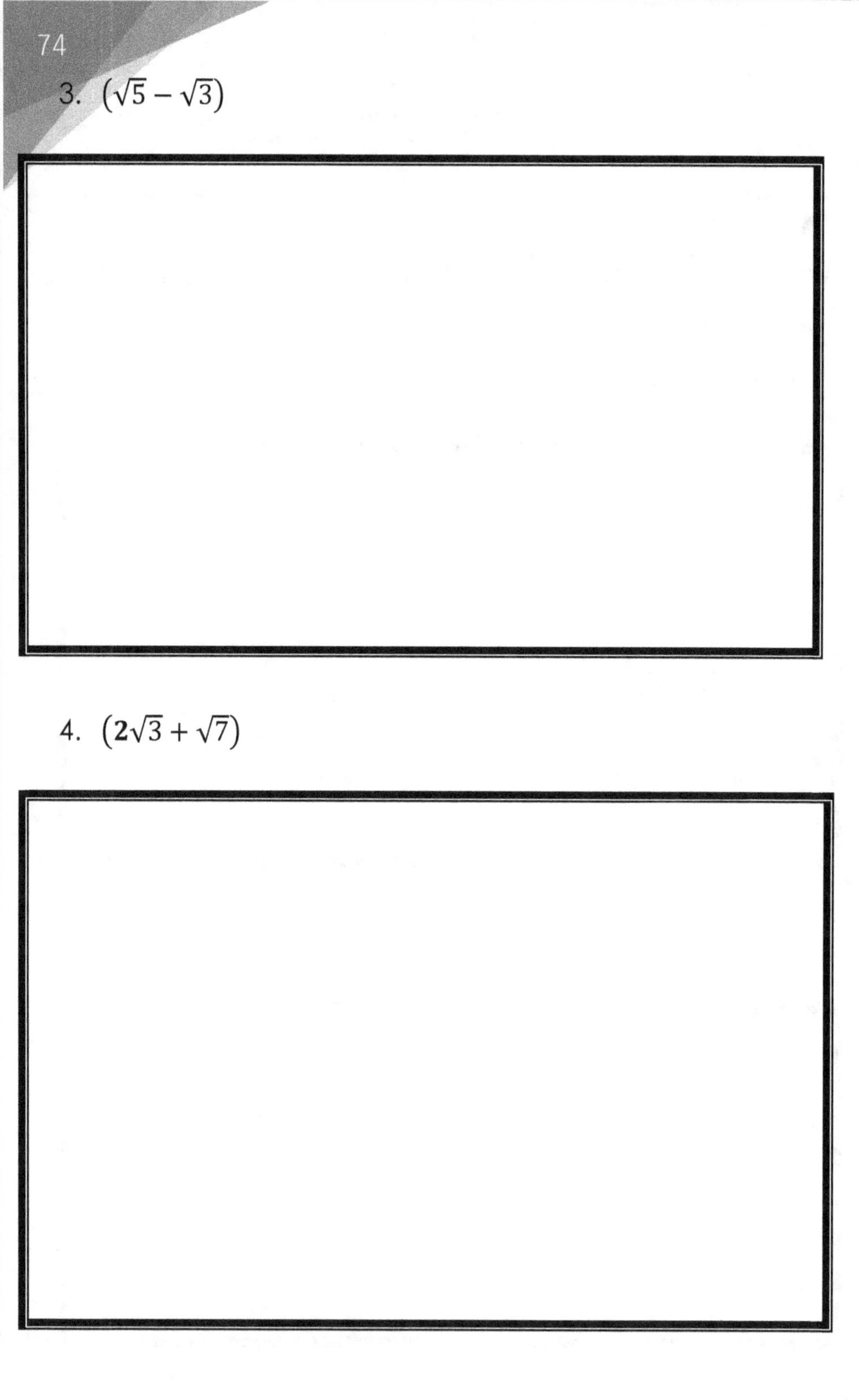

4. $\left(2\sqrt{3} + \sqrt{7}\right)$

5. $\left(5 - \sqrt{7}\right)$

Exercise 6.2

Simplify the following by rationalising the denominator

1. $\dfrac{5}{\sqrt{18}}$

2. $\dfrac{7}{\sqrt{5}}$

3. $\dfrac{4}{\sqrt{3}}$

4. $\dfrac{11}{\sqrt{6}}$

5. $\dfrac{2\sqrt{3}}{\sqrt{5}}$

6. $\dfrac{4\sqrt{5}}{\sqrt{2}}$

7. $\sqrt{\dfrac{2}{3}}$

8. $\sqrt{\dfrac{11}{7}}$

9. $\dfrac{\sqrt{2}\times\sqrt{3}\times\sqrt{7}}{\sqrt{5}\times\sqrt{2}}$

10. $\dfrac{\sqrt{12}\times\sqrt{27}}{\sqrt{3}}$

10. $\dfrac{\sqrt{12}\times\sqrt{27}}{\sqrt{3}}$

Exercise 6.3

Simplify the following by rationalising the denominator with binomial surd

1. $\dfrac{1}{3-\sqrt{5}}$

2. $\dfrac{\sqrt{5}+\sqrt{7}}{\sqrt{3}-2}$

2. $\dfrac{\sqrt{5}+\sqrt{7}}{\sqrt{3}-2}$

3. 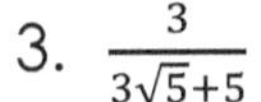 $\dfrac{3}{3\sqrt{5}+5}$

4. $\dfrac{2\sqrt{3}-\sqrt{7}}{2\sqrt{3}+\sqrt{7}}$

5. $\dfrac{2\sqrt{3}-\sqrt{5}}{3\sqrt{5}-5\sqrt{3}}$

6. $\dfrac{\sqrt{5}+3}{4-\sqrt{10}}$

7. $\dfrac{9+\sqrt{7}}{9-\sqrt{7}}$

8. $\dfrac{7\sqrt{3}-2\sqrt{2}}{1+\sqrt{3}}$

8. $\dfrac{7\sqrt{3}-2\sqrt{2}}{1+\sqrt{3}}$

9. $\dfrac{7}{2\sqrt{3}+2\sqrt{5}}$

10. $\dfrac{7\sqrt{5}-3\sqrt{2}}{2\sqrt{3}+3\sqrt{7}}$

Chapter 7

EQUALITY OF SURDS

The concept of equality f surd make us to determine the squareroot of a surdic number.

Square root of surds

To evaluate the square root of surds, study the illustration below.

$$\left(\sqrt{a}+\sqrt{b}\right)^2 = \left(\sqrt{a}+\sqrt{b}\right)\left(\sqrt{a}+\sqrt{b}\right)$$

$$= \left(\sqrt{a}\times\sqrt{a}\right) + \left(\sqrt{a}\times\sqrt{b}\right) + \left(\sqrt{b}\times\sqrt{a}\right) + \left(\sqrt{b}\times\sqrt{b}\right)$$

$$= \left(\sqrt{a\times a}\right) + \left(\sqrt{a\times b}\right) + \left(\sqrt{b\times a}\right) + \left(\sqrt{b\times b}\right)$$

$$= (a) + \left(\sqrt{ab}\right) + \left(\sqrt{ab}\right) + (b)$$

$$= (a) + (b) + \left(\sqrt{ab}\right) + \left(\sqrt{ab}\right)$$

$$= a + b + 2\left(\sqrt{ab}\right)$$

Therefore

$$\left(\sqrt{a}+\sqrt{b}\right)^2 = a + b + 2\left(\sqrt{ab}\right)$$

where (a

+ b) is a rational number and $2\left(\sqrt{ab}\right)$ ***is a surd expression***

Note

$$\sqrt{\left(\sqrt{a}+\sqrt{b}\right)^2} = \sqrt{a}+\sqrt{b}$$

Question 1 : Find the square root of $7 + 2\sqrt{10}$

Recall

$$\left(\sqrt{a}+\sqrt{b}\right)^2 = a + b + 2\left(\sqrt{ab}\right)$$

and

$$\sqrt{\left(\sqrt{a}+\sqrt{b}\right)^2} = \sqrt{a}+\sqrt{b}$$

If

$$\sqrt{7+2\sqrt{10}} = \sqrt{a}+\sqrt{b}$$

Therefore;

$$\left(\sqrt{a}+\sqrt{b}\right)^2 = 7+2\sqrt{10}$$

$$a+b+2\left(\sqrt{ab}\right) = 7+2\sqrt{10}$$

Compare the L.H.S surd expression with the R.H.S

$$a+b = 7 -----------(1)$$

$$2\left(\sqrt{ab}\right) = 2\sqrt{10}$$

$$ab = 10 ---------(2)$$

from equation (1)

$$a = 7-b --------(3)$$

substitute $\quad 7-b$ for a in equation (2)

In equation (2)

$$(7-b)b = 10$$

$$-b^2 + 7b = 10$$

$$b^2 - 7b + 10 = 0$$

Solve the quadratic Equation

$$b^2 - 7b + 10 = 0$$

Find two factors that, when summed, gives -7 (the coefficient of b) and when multiplied, it gives the product of the coefficient of b^2 (1) and$+10$, which equals $+10$.

$$Sum = -7 \qquad\qquad product = +10$$

$$-5 - 2 = -7 \qquad\qquad -5 \times (-2) = +10$$

The two factors are **−5** and **−2**, substitute **−5b − 2b** for **−7b** in the equation.

$$b^2 - 5b - 2b + 10 = 0$$

$$b(b - 5) - 2(b - 5) = 0$$

$$(b - 2)(b - 5) = 0$$

$$b - 2 = 0 \ or \ b - 5 = 0$$

$$b = 2 \ or \ 5$$

From equation (3)

$$a = 7 - b - - - - - - - (3)$$

$$When \quad b = 2, a = 5$$

$$b = 5, a = 2$$

Hence

The square root of $7 + 2\sqrt{10} = \sqrt{5} + \sqrt{2}$ (*twice*)

$$\sqrt{7 + 2\sqrt{10}} = \sqrt{5} + \sqrt{2} \ (\textbf{\textit{twice}})$$

Question 2 : Find the square root of $14 - 4\sqrt{6}$

Recall

$$\left(\sqrt{a} + \sqrt{b}\right)^2 = a + b + 2\left(\sqrt{ab}\right)$$

and

$$\sqrt{\left(\sqrt{a} + \sqrt{b}\right)^2} = \sqrt{a} + \sqrt{b}$$

If

$$\sqrt{14 - 4\sqrt{6}} = \sqrt{a} + \sqrt{b}$$

Therefore;

$$\left(\sqrt{a} + \sqrt{b}\right)^2 = 14 - 4\sqrt{6}$$

$$a + b + 2\left(\sqrt{ab}\right) = 14 - 4\sqrt{6}$$

Compare the L.H.S surd expression with the R.H.S

$$a + b = 14 - - - - - - - - - - (1)$$

$$2\left(\sqrt{ab}\right) = -4\sqrt{6}$$

Divide through by 2

$$\frac{2\left(\sqrt{ab}\right)}{2} = \frac{-4\sqrt{6}}{2}$$

$$\left(\sqrt{ab}\right) = -2\sqrt{6}$$

Square both sides

$$\left(\sqrt{ab}\right)^2 = \left(-2\sqrt{6}\right)^2$$

$$(ab) = 24$$

$$ab = 24 - - - - - - - - -(2)$$

from equation (1)

$$a = 14 - b - - - - - - - - (3)$$

substitute $14 - b$ for **a** in equation (2)

In equation (2)

$$(14 - b)b = 24$$

$$-b^2 + 14b = 24$$

$$b^2 - 14b + 24 = 0$$

Solve the quadratic Equation

$$b^2 - 14b + 24 = 0$$

Find two factors that, when summed, gives -14 (the coefficient of b) and when multiplied, it gives the product of the coefficient of b^2 (1) and $+24$, which equals $+24$.

$$Sum = -14 \qquad\qquad product = +24$$

$$-12 - 2 = -14 \qquad\qquad -12 \times (-2) = +24$$

The two factors are -12 and -2, substitute $-12b - 2b$ for $-14b$ in the equation.

$$b^2 - 12b - 2b + 24 = 0$$

$$b(b - 12) - 2(b - 12) = 0$$

$$(b - 2)(b - 12) = 0$$

$$b - 2 = 0 \; or \; b - 5 = 0$$

$$b = 2 \; or \; 12$$

From equation (3)

$$a = 14 - b - - - - - - - (3)$$

$$When \quad b = 2, a = 12$$

$$b = 12, a = 2$$

Hence

The square root of $14 - 4\sqrt{10} = \sqrt{12} - \sqrt{2} \; or \; \sqrt{2} - \sqrt{12}$

$$\sqrt{12} - \sqrt{2} = 2\sqrt{3} - \sqrt{2}$$

$$\sqrt{14 - 4\sqrt{6}} = 2\sqrt{3} - \sqrt{2} \; or \; \sqrt{2} - 2\sqrt{3}$$

Exercise 7.1

Find the squareroot of each of the following binomial surd.

1. $30 + 12\sqrt{6}$

2. $8 - 2\sqrt{15}$

3. $12 - 2\sqrt{35}$

3. $12 - 2\sqrt{35}$

4. $31 - 4\sqrt{21}$

Chapter 8

EQUATION SOLUTION ON SURD FORM

In this chapter we shall consider equation of surd having a variable in a single and an irrational form. The questions below will explains the process of solving a surd equations with variables.

Question 1: Solve $2\sqrt{3x+4} + x = 36$

$$2\sqrt{3x+4} + x = 36$$

Separate the terms with radicals

$$2\sqrt{3x+4} = 36 - x$$

Square both the L.H.S and R.H.S

$$\left(2\sqrt{3x+4}\right)^2 = (36-x)^2$$

$$4(3x+4) = (36-x)(36-x)$$

$$4(3x+4) = 1296 - 36x - 36x - x^2$$

$$(12x+16) = 1296 - 72x + x^2$$

$$x^2 - 72x - 12x - 16 + 1296 = 0$$

$$x^2 - 84x + 1280 = 0$$

$$x^2 - 64x - 20x + 1280 = 0$$

$$x(x-64) - 20(x-64) = 0$$

$$(x-64)(x-20) = 0$$

$$x = 64, x = 20$$

Since there are two answers for x, lets check for the right value that satisfy the equation.

When $x = 64$

$$2\sqrt{3(64) + 4} + 64 = 36$$

$$2\sqrt{192 + 4} + 64 = 36$$

$$2\sqrt{196} + 64 = 36$$

$$2(14) + 64 = 36$$

$$28 + 64 \neq 36$$

When $x = 64$ the equation is not satisfied.

When $x = 20$

$$2\sqrt{3(20) + 4} + 20 = 36$$

$$2\sqrt{60 + 4} + 20 = 36$$

$$2\sqrt{64} + 20 = 36$$

$$2(8) + 20 = 36$$

$$16 + 20 = 36$$

When $x = 20$ the equation is satisfied.

Hence $x = 20$

Question 2: Solve $\sqrt{x + 1} + \sqrt{x + 8} = 7$

$$\sqrt{x + 1} + \sqrt{x + 8} = 7$$

Square both the L.H.S and R.H.S

$$\left(\sqrt{x + 1} + \sqrt{x + 8}\right)^2 = 7^2$$

$$\left(\sqrt{x + 1} + \sqrt{x + 8}\right)\left(\sqrt{x + 1} + \sqrt{x + 8}\right) = 49$$

$$x + 1 + 2\sqrt{(x + 8)(x + 1)} + x + 8 = 49$$

$$x + x + 8 + 1 + 2\sqrt{(x + 8)(x + 1)} = 49$$

$$2x + 9 + 2\sqrt{(x+8)(x+1)} = 49$$

$$2x + 2\sqrt{(x+8)(x+1)} = 49 - 9$$

$$2\sqrt{(x+8)(x+1)} = 40 - 2x$$

Divide through by 2

$$\frac{2\sqrt{(x+8)(x+1)}}{2} = \frac{40 - 2x}{2}$$

$$\sqrt{(x+8)(x+1)} = 20 - x$$

Square both side

$$(x+8)(x+1) = (20 - x)^2$$

$$x^2 + x + 8x + 8 = 400 - 20x - 20x + x^2$$

$$x^2 + 9x + 8 = 400 - 40x + x^2$$

$$x^2 - x^2 + 9x + 40x - 400 + 8 = 0$$

$$+9x + 40x - 400 + 8 = 0$$

$$49x - 392 = 0$$

$$49x = 392$$

Divide through by 49

$$\frac{49x}{49} = \frac{392}{49}$$

$$\boldsymbol{x = 8}$$

Question 3 : Solve $x + \sqrt{x + 5} = 7$

$$x + \sqrt{x + 5} = 7$$

$$\sqrt{x + 5} = 7 - x$$

Square both the L.H.S and R.H.S

$$\left(\sqrt{x + 5}\right)^2 = (7 - x)^2$$

$$x + 5 = 49 - 7x - 7x + x^2$$

$$x + 5 = 49 - 14x + x^2$$

$$49 - 5 - 14x - x + x^2 = 0$$

$$x^2 - 15x + 44 = 0$$

Find two factors that, when summed, gives **−15** (the coefficient of **x**) and when multiplied, it gives the product of the coefficient of x^2 (**1**) and **+44**, which equals **+44**.

$$\textbf{Sum } = \textbf{−15} \qquad\qquad \textbf{product } = \textbf{+44}$$

$$-11 - 4 = -15 \qquad\qquad -11 \times (-4) = +44$$

The two factors are **−11** and **−4**, substitute **−11x − 4x** for **−14x** in the equation.

$$x^2 - 11x - 4x + 44 = 0$$

$$x(x - 11) - 4(x - 11) = 0$$

$$(x - 4)(x - 11) = 0$$

$$x - 4 = 0 \ or \ x - 11 = 0$$

$$x = \textbf{4 or 11}$$

Since there are two answers for x, lets check for the right value that satisfy the equation.

When $x = 4$

$$4 + \sqrt{4 + 5} = 7$$

$$4 + \sqrt{9} = 7$$

$$4 + 3 = 7$$

When $x = 4$ the equation is satisfied.

When $x = 11$

$$11 + \sqrt{11 + 5} = 7$$

$$11 + \sqrt{16} = 7$$

$$11 + 4 \neq 7$$

When $x = 11$ the equation is not satisfied.

Hence $x = 4$

Question 4: Solve $\sqrt{x + 5} + \sqrt{x} = 5$

$$\sqrt{x + 5} + \sqrt{x} = 5$$

Square both the L.H.S and R.H.S

$$\left(\sqrt{x + 5} + \sqrt{x}\right)^2 = 5^2$$

$$\left(\sqrt{x + 5} + \sqrt{x}\right)\left(\sqrt{x + 5} + \sqrt{x}\right) = 25$$

$$x + 5 + 2\sqrt{(x + 5)x} + x = 25$$

$$x + x + 5 + 2\sqrt{(x + 5)x} = 25$$

$$2x + 5 + 2\sqrt{(x + 5)x} = 25$$

$$2\sqrt{(x + 5)x} = 25 - 5 - 2x$$

$$2\sqrt{(x + 5)x} = 20 - 2x$$

Divide through by 2

$$\sqrt{(x + 5)x} = 10 - x$$

Square both side

$$(x + 5)(x) = (10 - x)^2$$

$$x^2 + 5x = 100 - 10x - 10x + 4x^2$$

$$x^2 + 5x = 100 - 20x + x^2$$

$$x^2 - x^2 - 20x - 5x + 100 = 0$$

$$-25x + 100 = 0$$

$$-25x = 0 - 100$$

$$-25x = -100$$

Divide through by -25

$$\frac{-25x}{-25} = \frac{-100}{-25}$$

$$\boldsymbol{x = 4}$$

Question 5: Solve $2\sqrt{3x+1} + 3\sqrt{4x+5} = 23$

$$2\sqrt{3x+1} + 3\sqrt{4x+5} = 23$$

Square both the L.H.S and R.H.S

$$\left(2\sqrt{3x+1} + 3\sqrt{4x+5}\right)^2 = 23^2$$

$$\left(2\sqrt{3x+1} + 3\sqrt{4x+5}\right)\left(2\sqrt{3x+1} + 3\sqrt{4x+5}\right) = 529$$

$$4(3x+1) + 2(3 \times 2)\sqrt{(4x+5)(3x+1)} + 9(4x+5) = 529$$

$$12x + 4 + 12\sqrt{(4x+5)(3x+1)} + 36x + 45 = 529$$

$$12x + 36x + 4 + 45 + 12\sqrt{(4x+5)(3x+1)} = 529$$

$$48x + 49 + 12\sqrt{(4x+5)(3x+1)} = 529$$

$$12\sqrt{(4x+5)(3x+1)} = 529 - 49 - 48x$$

$$12\sqrt{(4x+5)(3x+1)} = 480 - 48x$$

Divide through by 12

$$\frac{12\sqrt{(4x+5)(3x+1)}}{12} = \frac{480 - 48x}{12}$$

$$\sqrt{(4x+5)(3x+1)} = 40 - 4x$$

Square both side

$$(4x+5)(3x+1) = (40 - 4x)^2$$

$$12x^2 + 4x + 15x + 5 = 1600 - 160x - 160x + 16x^2$$

$$12x^2 + 19x + 5 = 1600 - 320x + 16x^2$$

$$16x^2 - 12x^2 - 320x - 19x + 1600 - 5 = 0$$

$$\mathbf{4x^2 - 339x + 1595 = 0}$$

Find two factors that, when summed, gives **−339** (the coefficient of x) and when multiplied, it gives the product of the coefficient of x^2 (**4**) and **+1595**, which equals **+6380**.

Sum $= -339$ *product* $= +6380$

$-20 - 319 = -339$ $-20 \times (-319) = +6380$

The two factors are **−20** and **−319**, substitute **−20x − 319x** for **−339x** in the equation.

$$4x^2 - 20x - 319x + 1595 = 0$$

$$4x(x - 5) - 319(x - 5) = 0$$

$$(x - 5)(4x - 319) = 0$$

$$x - 5 = 0 \text{ or } 4x - 319 = 0$$

$$x = 5 \text{ or } \frac{319}{4}$$

Since there are two answers for x, lets check for the right value that satisfy the equation.

When $x = 5$

$$2\sqrt{3(5) + 1} + 3\sqrt{4(5) + 5} = 23$$

$$2\sqrt{16} + 3\sqrt{25} = 23$$

$$(2 \times 4) + (3 \times 5) = 23$$

$$8 + 15 = 23$$

When $x = 5$ the equation is satisfied.

When $x = \frac{319}{4}$

$$2\sqrt{3\left(\frac{319}{4}\right)+1}+3\sqrt{4\left(\frac{319}{4}\right)+5} \neq 23$$

When $x = \frac{319}{4}$ the equation is not satisfied.

Hence $x = 5$

Exercise 8.1

Solve the following.

   1. $2\sqrt{x+1}+x=7$

2. $3\sqrt{x + 2} - x = 4$

3. $\sqrt{x+5} + \sqrt{x} = 5$

4. $\sqrt{5x - 4} + \sqrt{x} = 6$

5. $5x - 2\sqrt{2x + 3} = 9$

ANSWERS

Exercise 2.1

1. $6\sqrt{2}$
2. $7\sqrt{3} - 3\sqrt{2}$
3. $15\sqrt{5}$
4. $10\sqrt{7}$
5. $10\sqrt{11}$
6. $630\sqrt{3}$
7. 12
8. $875\sqrt{7}$
9. $96\sqrt{30}$
10. $7\sqrt{5} + 16\sqrt{3}$

Exercise 3.1

1. 6
2. 2496
3. $\frac{13}{12}$
4. 4
5. $5\sqrt{2} - 9\sqrt{3}$
6. $9 + 4\sqrt{5}$
7. $10 + 2\sqrt{21}$
8. 1
9. $\frac{\sqrt{3}}{2}$
10. 7

Exercise 4.1

1. $4\sqrt{3}$
2. $2\sqrt{5}$
3. $5\sqrt{2}$
4. $3\sqrt{11}$
5. $5\sqrt{6}$
6. $7\sqrt{2}$
7. $5\sqrt{3}$
8. $18\sqrt{2}$
9. $20\sqrt{2}$
10. $9\sqrt{3}$

Exercise 4.2

1. $\sqrt{448}$
2. $\sqrt{200}$
3. $\sqrt{363}$
4. $\sqrt{245}$
5. $\sqrt{891}$
6. $\sqrt{200}$
7. $\sqrt{864}$
8. $\sqrt{810}$
9. $\sqrt{50}$
10. $\sqrt{432}$

Exercise 5.1

1. $11\sqrt{3}$
2. $-5\sqrt{2}$
3. $33\sqrt{2}$
4. $4\sqrt{7}$
5. $-\sqrt{13}$
6. $-\sqrt{14}$
7. $42\sqrt{2}$

Exercise 5.2

1. 4.242
2. 6.323
3. 13.86
4. 12.07
5. 11.18

Exercise 5.3

1. $7\sqrt{2} - 5\sqrt{3}$
2. $12\sqrt{6} + 8\sqrt{15} + 12 + 4\sqrt{10}$
3. $30 - 4\sqrt{10}$
4. $11 - \sqrt{21}$
5. $5\left(9 - 4\sqrt{2}\right)$
6. $24\sqrt{10}$
7. 44
8. 60
9. $6\sqrt{5}$
10. $2\sqrt{3} + 6$

Exercise 6.1

1. 29
2. 7
3. 2
4. 5
5. 18

Exercise 6.2

1. $\dfrac{5\sqrt{18}}{18}$
2. $\dfrac{7\sqrt{5}}{5}$
3. $\dfrac{4\sqrt{3}}{3}$
4. $\dfrac{11\sqrt{6}}{6}$
5. $\dfrac{2\sqrt{15}}{5}$
6. $2\sqrt{10}$
7. $\dfrac{\sqrt{6}}{3}$
8. $\dfrac{\sqrt{77}}{7}$
9. $\dfrac{\sqrt{105}}{5}$
10. $6\sqrt{3}$

Exercise 6.3

1. $\dfrac{3+\sqrt{5}}{4}$

2. $-\sqrt{15} - 2\sqrt{5} - \sqrt{21} - 2\sqrt{7}$

3. $\dfrac{9\sqrt{5}-15}{20}$

4. $\dfrac{19-4\sqrt{21}}{5}$

5. $\dfrac{-\sqrt{15}-15}{30}$

6. $\dfrac{4\sqrt{5}+5\sqrt{2}+4\sqrt{3}+\sqrt{30}}{6}$

7. $\dfrac{88+18\sqrt{7}}{74}$

8. $\dfrac{7\sqrt{3}-2\sqrt{2}+2\sqrt{6}-21}{2}$

9. $\dfrac{-7(\sqrt{3}-\sqrt{5})}{4}$

10. $\dfrac{-(14\sqrt{15}-21\sqrt{35}-6\sqrt{6}+9\sqrt{14})}{51}$

Exercise 7.1

1. $\sqrt{7} + \sqrt{5}$

2. $3\sqrt{2} + 2\sqrt{3}$

3. $\sqrt{3} - \sqrt{5},\ \sqrt{5} - \sqrt{3}$

4. $\pm\left(2\sqrt{7} - \sqrt{3}\right)$

Exercise 8.1

1. 3
2. −1,2
3. 4
4. 4
5. 3

Improve your Math Skills with other Books

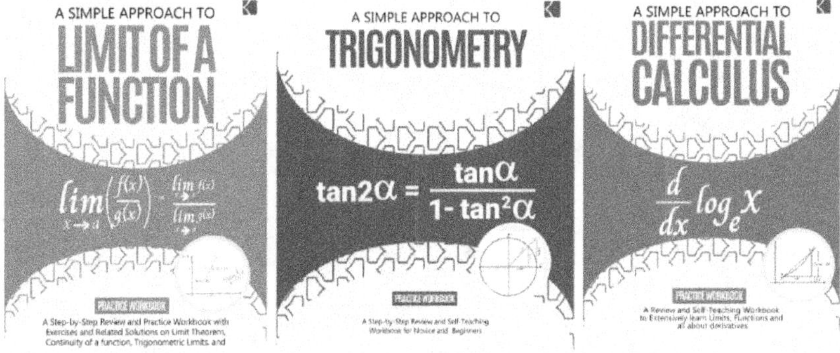